ACCESS TO MATH

FORMULAS

TEACHER'S MANUAL

GLOBE FEARON EDUCATIONAL PUBLISHER
A Division of Simon & Schuster
Upper Saddle River, New Jersey

ACCESS TO MATH

Contents

Meeting Student Needs 1

    ESL/LEP Strategies 1

    Cooperative Learning 1

    Manipulatives 2

Lesson Plans 3

Answer Key 19

MEETING STUDENT NEEDS

Most classrooms include students with a wide variety of different needs and learning styles. As you plan your lessons, you can include teaching strategies that address the needs you have identified. Although the ideas that follow are stated in terms of specific needs, all students can benefit from these strategies.

ESL/LEP Strategies

Mathematical learning is made up of three elements: the mathematical concept, its symbolic notation, and the vocabulary with which these elements are expressed. The ESL student will benefit from a constant focus on the interrelationship among these elements. Whenever you present a mathematical concept, stress the connection between the language and the concept by giving simply worded definitions and specific examples. When you present a symbolic notation, stress the connection between the language and the symbols both orally and in written form. Since this approach can be helpful to native speakers also, it will be advantageous to the entire class.

A variety of techniques can help strengthen students' math and language skills.

- Assign ESL students to a small group within which they can feel comfortable and receive individual attention. If there are several students whose first language is the same, assign them in pairs to a group. Add a native speaker of English who wants to help others and who will be patient and understanding. Alternatively, partner ESL students with native speakers at all stages of learning, from the conceptual stage to practice exercises.
- Use a combination of written and spoken representations of terms and mathematical symbols. Always connect spoken terms with what they look like in writing, and vice versa.
- Regularly write essential vocabulary on the board and say these words aloud. Also ask students to say and write the words.
- When necessary, rephrase definitions of vocabulary or math terms in several ways. If students cannot grasp the definition phrased in one form, it may reach them more efficiently when worded another way.
- Encourage students to paraphrase definitions in their own words. This increases language facility and also lets students restate the ideas in terms with which they are comfortable.
- Continually give specific examples of concepts or terms. Whenever possible, use examples from everyday life. Draw on students' previous knowledge and experience. Also ask students to give their own examples.
- Whenever necessary, model concepts and processes with concrete materials or with diagrams. Also encourage students to use manipulatives and diagrams.
- Have students create and use flash cards that feature vocabulary words, math symbols, and other useful mathematical elements. Set aside time for flash card work on a regular basis.

Cooperative Learning

Success with cooperative learning groups depends on teaching students, in incremental steps, how to work together. Employers consider the ability to solve problems as part of a team a key skill in the people they hire. Additionally, many at-risk students learn best in a cooperative learning setting. These two facts alone make it worthwhile to spend time teaching students how to work effectively and stay on task in a group situation.

Begin by familiarizing students with the actual process of forming groups quickly and efficiently. Start with partners or groups of three, of mixed abilities. Teach students how to move desks or chairs quickly and quietly. Insure individual participation by assigning specific roles within the interdependent group. For example, roles might include summarizer, checker, reader, researcher, recorder, and so forth. As students become more familiar with the process, they can often choose, assign, and title their own roles.

Assign a mathematical task to be accomplished by the group. Explain that the group will have accomplished the task when each member can answer four out of five problems correctly. Allow plenty of time and explain that finishing first among group members (a competitive rather than a cooperative approach) is not the point. Rather, the point is for the group to work together to ensure that all members master the skill. You may want to test mastery at a second class meeting, after a short review. If a group does not meet the criteria, ask members to work together to clear up misconceptions and confusions. Encourage those who have succeeded to coach and motivate others who have not yet been successful.

Some students may have difficulty buying into group work and group accomplishments. Tell

MEETING STUDENT NEEDS

them that teamwork is a realistic reflection of the way work is done in the world outside the classroom. Help quicker students to realize that the best understanding of a concept comes from teaching it to someone else. Assure them that they are internalizing their own knowledge into long-term memory when they explain a concept to another student.

The following tips will help you create a beneficial environment for learning math cooperatively.

- Allow the task to determine group size. Groups of 2 or 3 might work to understand or review a process such as dividing fractions. Groups of up to 6 might formulate problems or work on solving complex word problems.
- Keep group meetings and tasks short at first and gradually lengthen them.
- Before work begins, give students a clear statement of what they need to know and do to complete the assignment. This information should be written on the chalkboard or on handouts. Include the following:
  a. mathematical objectives (what the group will learn, what skills and processes the group will use, and what product or solution the group should expect to arrive at)
  b. collaborative objectives (how the group will function to achieve its goals)
  c. the amount of time that will be given to complete the assignment
  d. the criteria by which the work will be evaluated (for example, will the group have to explain the process it used? Come up with a single, accurate answer? Give a range of possible answers?)
- Assign necessary roles, such as recorder or checker, at the start of the activity.
- Monitor the groups' progress. When necessary, intervene to answer questions, reinforce math skills, or stress appropriate group procedures.
- Talk about and teach conflict management skills to help students resolve differences. Remind them that they will often have to work with people they did not choose, but that they can still get the job done.
- Assess results in terms of both groups and individuals. When possible, have groups compare their results with those of other groups.

MANIPULATIVES

The term *manipulatives* is used to describe any objects students can handle and work with as they learn mathematics. Multiplication grids, chips, counters, and coins are all useful manipulatives for many mathematical concepts.

All students learn better and remember longer when they have concrete experiences that relate to abstract ideas. Research tells us that at-risk students have a higher proportion of tactile and kinesthetic learning styles than the student population in general. Thus, these students learn best if they are introduced to a topic in ways that involve touching and moving objects (tactile) and moving their large muscles (kinesthetic).

Manipulatives can be used to demonstrate concepts and processes in many areas of intermediate and advanced classrooms, as they are in primary grades. Of course, adjustments should be made for the different situation. For instance, the age level of the students should be considered when choosing materials. Also, because upper-level classrooms might not have budget allocations for "official" manipulatives such as place value blocks, everyday materials such as paper clips or even liquids can be used. Here are some tips for working with manipulatives:

- For many situations, uniform size and shape is essential. Paper clips may be useful. Paper is easily obtained and can be cut into any size and shape. Dried beans and pasta are inexpensive even in large quantities. (However, the use of food items as manipulatives may be frowned on in certain communities, such as some Hispanic communities, as wasteful.)
- In some situations, color coding is helpful. Marbles of different colors make good choices.
- In situations where parts of a whole must be modeled and perhaps reshaped or removed, consider modeling clay, kneaded erasers, dough made of flour, salt, and water.
- Measuring implements such as marked measuring cups may be useful for investigating fractions or decimals.
- A balance can be useful for more than just measurement activities. The concept of making sure that both sides of an equation are equal in value can be demonstrated with pennies and a balance. An unknown quantity can be shown with pennies wrapped in paper. (Caution: New pennies and old pennies have slightly different weights, so pre-weigh pennies to be sure they are uniform.)
- Play money can be useful for both money computation and place value work. Bills and coins can be made from paper and cardboard if premade play money is not available.
- Diagrams can be useful as a substitute for manipulative objects in some situations.
- Students themselves can become "manipulatives," acting out certain processes and substitutions.

LESSON PLANS

LESSON 1

Introducing the Lesson
Ask students to explain how they would determine the total cost of a movie for a group of children and adults if the children's tickets are $2 each and the adults' are $4 each.

Teaching the Lesson
On the chalkboard, write the symbols: $+$, $-$, $\times$, $\div$, and $\sqrt{\ }$. Ask students to describe this set (they are mathematical symbols). Then, write these two columns with the given labels:

Numerical Expression	Value
$3 + 7$	10
$5 \times 3 - 9$	6
$44 \div 4$	11

Tell students that numerical expressions are combinations of numbers and mathematical operations. Ask students to look at the value column and tell in their own words how to find the value of a numerical expression.

When reading through the lesson with students, be sure they understand the meaning of the word *variable*. Ask them to give examples of variables that they define.

After reading and working out the given example, ask students how much money would be raised if they sold 90 adult tickets and 50 children's tickets [$(5 \times 90) + (3 \times 50) = 450 + 150 = \$600$]

Error Analysis
When writing what a variable represents, some students may not be specific enough. For example, in Exercise 4, they may write p = plums. Instead of writing p = the number of pounds of plums. Tell students to be sure to always include a label—such as number of pounds, or number of dollars, or cost of jacket—in the statement of what the variable represents.

Follow-Up
CONSUMER CONNECTION Ask students what the prices are of an adult movie ticket and a child's movie ticket in their area or at the theater they go to most. Have them write an expression that will give the total cost of a adult tickets and c children's tickets using that information.

LESSON 2

Introducing the Lesson
Write on the chalkboard: $2 + 8 \times 6$. Ask students to compute the answer. Some may say 60 and some may say 50. Tell students that in this lesson they will learn how to tell which answer is correct.

Teaching the Lesson
Students may find this topic difficult, especially when the expressions they must work with have many terms. It will help them to go through many different kinds of worked-out examples. You may want to have students work out each of the following:

a. $3 \times (9 - 5) - 6$
$3 \times 4 - 6$
$12 - 6 = 6$

b. $4^2 - 2 \times 5$
$16 - 2 \times 5$
$16 - 10 = 6$

c. $2^3 + 12 \div 4$
$8 + 12 \div 4$
$8 + 3 = 11$

d. $20 - 8 \div 2$
$20 - 4 = 16$

As needed, review the meaning of exponents.

Error Analysis
Some students may make errors because they lose

LESSON PLANS

track of where they are in finding the value of the whole expression. For example, when finding the value of (30 − 9) × (6 + 17), they may do the first parentheses, 30 − 9 = 21, and stop there. Encourage students to rewrite the entire expression in its revised form each time they do one part of it.

Follow-Up

REAL-LIFE CONNECTION Ask students to describe three everyday tasks that must be done in a certain order and three that do not have to be done in a certain order. For example, you must put on socks before shoes but you can put ingredients for muffins in a bowl in any order.

LESSON 3

Introducing the Lesson

Write these expressions on the chalkboard and ask students to translate each into a word phrase:

$a + 7$ (add 7 to a)

$4b$ (multiply b by 4)

$c - 10$ (subtract 10 from c)

$10 - d$ (subtract d from 10)

$5x - 1$ (multiply 5 times x, then subtract 1)

Teaching the Lesson

In this lesson, students put together what they learned about variables and expressions in Lesson 1 with order of operations from Lesson 2. Point out that an expression like $\$5 + \$3 \times (h - 1)$ will have a different value for every different value of h that is used. When h is replaced by 6, you can find one value of the expression. Emphasize that within an expression, each variable represents only one value.

It may help students to see the expression in the lesson evaluated for $h = 2$ ($\$8$).

Error Analysis

Some students may get $64.50 for Exercise 7. If so, they forgot the correct order of operations. The correct order is: $\$1.50 + \$20d = \$1.50 + \$20 \times 3 = \$1.50 + \$60 = \$61.50$. You can use a mnemonic device to help students remember the order of operations.

Follow-Up

FAMILY MATH Encourage students who sometimes care for younger children to play store using play money. They can ask the children questions such as: How much should I pay for 2 stickers that cost 5 cents each and 3 markers that cost 10 cents each? These students can record children's responses as algebraic expressions.

LESSON 4

Introducing the Lesson

Many actions have an inverse action. An inverse action undoes the original action. For example, the action of tying shoes has the inverse action of untying shoes. Ask students to give you the inverse actions for each of the following: going three steps forward (going three steps backward), sitting down (standing up), writing a word (erasing a word).

Teaching the Lesson

Start by discussing the expression $x + 80$. Ask students to tell what the value of this expression is when $x = 10$ (90), when $x = 40$ (120) and when $x = 0$ (80). Then write $x + 80 = 200$ on the chalkboard. Point out that this statement is an equation. To solve the equation, they must find the value of x that makes it true.

The equations in the lesson can be solved in one step. Students must determine the inverse operation to use in each case. When students have found the solution, encourage them to check their answer in the original equation. For example, let $x = 120$ in the equation $x + 80 = 200$. Then $120 + 80 = 200$, $200 = 200$ and it checks.

4 FORMULAS

Some students may prefer to arrange their work vertically for equations that involve addition or subtraction. For example, they may write:

$$\begin{array}{rl} x + 80 = & 200 \\ -80 & -80 \\ \hline x = & 120 \end{array}$$

Error Analysis
Students may not know what step to take to solve an equation because they do not understand inverse operations. Have students work with a calculator, starting with one number, doing an operation, and then doing the inverse operation to get back to the original number.

Follow-Up
CROSS-CURRICULAR CONNECTION: HISTORY Students may be interested to know that the word *algebra* is a Latin variant of the Arabic word *al-jabr*, taken from the title of a book written in Baghdad in about 825 A.D. by the Arab mathematician Mohammed ibn-Musa al-Khowarizmi. This book tells how to solve equations. Interested students can visit the school library to find out more about the history of mathematical topics. Solicit the help of your school librarian.

s by using the same principle of inverse operations that was used in the previous lesson.

The Guided Practice and Exercises ask that students solve the formulas for a different variable. Give specific real-life numerical examples some for the formulas. For example, a car going 50 miles per hour (rate) for 3 hours (time) will cover 150 miles (distance).

Some of these formulas will actually require two steps. To solve Exercise 6 for d, students must add d to each side first, then subtract p from each side. In Exercise 7, some may multiply each side by 2 first, then divide each side by b.

Error Analysis
Some students may have difficulty focusing on one of several variables in a formula. Have students rewrite the formula using a different color for the variable they must solve for.

Follow-Up
CONSUMER CONNECTION Ask students what they would need to know to determine the amount they will pay at a pizza parlor if they have a coupon for some money off the total bill. Ask them to write a formula for the amount they will pay, p, if the total bill is b and the coupon amount is c. ($p = b - c$)

LESSON 5

Introducing the Lesson
Ask students to tell you any formulas that they remember. If responses are few, ask questions such as: What is the formula for changing feet to yards? List any formulas that are named on the board and write what each letter in the formula represents.

Teaching the Lesson
Have a volunteer read the problem about Amiri's garden aloud. Draw a diagram on the chalkboard that illustrates the problem. Under the diagram, write the formula, $P = 4s$. The formula is solved for

LESSON 6

Introducing the Lesson
Ask students to tell about trips they have taken: how many miles they traveled, how long the trips took, what means of transportation they used, and how fast they were able to travel. Have them estimate how far it is possible to travel in one day using various modes of travel.

Teaching the Lesson
Problems about distance, rate, or time can all be solved using one formula, $d = rt$. In this lesson, students learn to recognize which variable they are

LESSON PLANS

looking for and rewrite the formula, if necessary, for that variable.

Notice that the label used for rate is the ratio *mi/h* and has the same meaning as the label *mph*, which may be more familiar to students.

You may want to discuss with students that the formula makes certain assumptions about travel. For example, it assumes that the rate of speed is always constant. Ask students to tell you if this is realistic and why or why not.

Discuss the idea that if a rate of speed is not constant, the *average* rate of speed can be used to make calculations about time and distance. Explain that if a vehicle traveled at various speeds over one hour's time and covered 50 miles, its average speed would be $50 mi/h$.

Error Analysis

Students may have trouble knowing which variable they are looking for. They may benefit from making a table with the column heads: Distance, Rate, and Time. Have them make three rows in the table. In each row, have them leave one column blank (a different column for each row) and write a problem for that row.

Follow-Up

REAL-LIFE CONNECTION Ask students how they would determine the time it would take them to walk 10 miles if they knew they could walk 3 miles in one hour (3 hours 20 minutes). Have them explain how they got their answer and which variation of the formula they used.

LESSON 7

Introducing the Lesson

Ask students to tell you what they know about the differences between the Fahrenheit temperature scale and the Celsius scale. Ask them to suggest a hot, cold, and comfortable temperature in degrees Celsius. Ask them to name local places where they have seen either or both scales displayed.

Teaching the Lesson

Draw a thermometer on the chalkboard. On the left side, mark a point near the top of the thermometer and label it 212°F and mark a point near the bottom and label it 32°F. On the right side, label the top point 100°C and the bottom point 0°C. Point out that the top temperature is the same amount of heat, the boiling point of water, and that the bottom temperature is the same amount of heat, the freezing point of water, but that they are labeled using two different scales.

Ask students to look at the thermometer you drew and answer these questions: How many Fahrenheit degrees are there between the freezing point and boiling point of water? (180) How many Celsius degrees are there between these two points? (100) Which degree is bigger, a Fahrenheit degree or a Celsius degree? (Celsius)

Ask students to rewrite the formulas using decimals in place of fractions. This can be useful when using a calculator. [F = 1.8C + 32; C = 0.56(F − 32), $\frac{5}{9} = 0.555...$]

Error Analysis

Some students may not follow order of operations rules when evaluating the formula. Remind students that operations within parentheses must be done first and multiplication and division should be done before addition or subtraction.

Follow-Up

CAREERCONNECTION Scientists often use the Celsius scale for reporting temperatures in their experiments. The Celsius scale was invented in 1742 by Anders Celsius, a Swedish astronomer. It is used in places that use the metric system and in all scientific work. Students may wish to research the Kelvin scale, which is also widely used by scientists.

6 FORMULAS

LESSON 8

Introducing the Lesson

Ask students why people put their money in a bank. Then ask why they may choose one bank over another. What are some benefits that banks provide?

Teaching the Lesson

The problems in this lesson are based on simple interest even though, in real life, most interest is compounded. To find the simple (that is, not compounded) interest, we use the formula $I = prt$ where I is the amount of interest, p is the principal or the amount invested, r is the rate per year expressed as a decimal or fraction and t is the time in years. The rate of interest is usually given as a percent and must be changed to a decimal or fraction for use in the formula.

The example in Guided Practice shows students what to do when any of the other variables — principal, rate, or time—must be found: the formula must be rewritten.

Error Analysis

Students may make errors in computing with the interest formula because they do not change the percent to a decimal or change it incorrectly. For example, students may use 0.3 for 3%. Remind these students to always move the decimal point two places to the left when changing a percent to a decimal. Guide students to use number sense to check that their answers make sense. Discuss a reasonable range of interest rates, pointing out that students should look back at their computations when rates are much higher that 10% or if the amount of interest is more than 10% of the principal.

Follow-Up

CONSUMER CONNECTION Invite students to investigate different interest rates and benefits offered by the banks in their community. Suggest that they prepare a chart that compares the interest earned by $100 in all the different banks.

LESSON 9

Introducing the Lesson

Ask students to list items that they buy that are subject to sales tax. Ask: What is the sales tax rate in your community?

Teaching the Lesson

Many states in the United States have a sales tax. In addition, some counties, towns, or cities levy an additional sales tax. The tax rate of percent can often include a mixed number, as in $2\frac{1}{2}\%$. Have students rewrite the fractional part of this kind of percent using a decimal first and then move the decimal point two places to the left for use in the formula. Therefore, $2\frac{1}{2}\%$ changes to 2.5%, which becomes 0.025 when computing the sales tax.

Point out that there are two formulas involved in sales tax problems. Multiplication is used to find the amount of the sales tax based on the price of the item, and addition is used to find the total cost. Caution students to read problems carefully to determine if the amount of sales tax is asked for or if the total cost including sales tax is called for.

Error Analysis

When using a calculator, students may see decimal answers shown to the tenths place. For example, in Exercise 2, the calculator display will read 1.2 for the sales tax on the sweatshirt. Some students may make an error when adding mentally to find the total cost because they ignore place value. For Exercise 2, they may say the total cost is $20.12 or $21.02. Have students who make this error write the sales tax using dollars and cents notation as soon as they find it, or have them use the calculator for both steps.

LESSON PLANS

Follow-Up

REAL-LIFE CONNECTION Tell students that the sales tax in Massachusetts is 5%. Ask them how much someone from that state would save if they bought a $10,000 car in a place that had a 2% sales tax. ($300)

LESSON 10

Introducing the Lesson

Ask students to tell you what kind of things they buy at a discount. Have a volunteer explain what the word means. Ask the students if there is a particular place that always gives a discount or if there is a time of year when they notice more discounts being offered. Have them discuss why they think stores offer discounts.

Teaching the Lesson

A discount is a reduction from the original price. Other names for original price are *list price* and *regular price*. Discounts are often given as percents but are sometimes listed as fractions. Some examples are 50% off or $\frac{1}{2}$ off.

To find the amount of the discount, students must multiply the original price by the discount rate as a decimal or fraction. Discuss with students the advantages of computing with the rate expressed as a fraction when compatible numbers are involved, as in Exercises 9 and 10. If the result has more than two decimal places, students must round to the hundredths place so that their answers are in dollars and cents. To find the final cost, they subtract the discount from the original price. As a check, students should note that the final cost should be less than the original price of the item.

Error Analysis

In Exercise 11, students may compare the numbers in the problem without regard for what they represent. Therefore, they may conclude that the best deal is the reduced price of $29.99. Point out that they must find the actual amount of discount in each case. They will see that $\frac{1}{3}$ off gives the greatest reduction in price.

Follow-Up

CONSUMER CONNECTION Have students check newspaper ads for sale announcements. Have them cut out or copy the ads and make up problems based on the information in the ads. Then, have them exchange their problems with other students for solving.

LESSON 11

Introducing the Lesson

Point to a window or door in the classroom that could be weatherstripped. Ask students how they would determine the length of weatherstripping that would be needed. Draw a diagram on the chalkboard and indicate which measurements would have to be taken and what to do with the measures.

Teaching the Lesson

Ask the students how many sides home plate has. Ask them to name other situations in which someone would need to know the perimeter of a figure. They may suggest trimming a table cloth, fencing in a yard, or framing a picture. Encourage students to name as many examples as they can think of.

In discussing the perimeter of a rectangle, ask students if they remember a formula or if they can see a shortcut. Since two pairs of sides are equal, they can use the formula $p = 2l + 2w$ or $p = 2(l + w)$.

For Exercises 7–10, encourage students to draw the figure described and label all of its sides so that it will be clear how to find the perimeter.

Error Analysis

In Exercise 6, some students may get the result 24 m. This means that they only added the lengths

8 FORMULAS

that were labeled. Remind students that the figure has six sides, so they must add six lengths. Guide them to use the hidden information the drawing provides to determine the lengths of the unlabeled sides.

Follow-Up

CONSUMER CONNECTION Ask students to determine how much wood trim they would need to place trim all around the ceiling of one of the rooms in their homes. Have them share the strategies they used to solve this problem.

LESSON 12

Introducing the Lesson

Draw a circle on the chalkboard. Ask students to tell you what they know about circles. Mark and label these parts as they are discussed: center, radius, diameter, and circumference. Ask students to estimate, just by looking, how many times greater than the diameter the circumference is.

Teaching the Lesson

You may want to begin with an experiment to find the value of π. Take coins or other circular objects. Have students measure the diameter of each in millimeters. Then, have them measure the circumference of each in millimeters. (They can roll the object along a sheet of paper or wrap string around it to find the circumference.) For each object, have them divide the circumference by the diameter, using a calculator. They should get numbers close to 3.14. Point out that this ratio of circumference to diameter is the same for any circle and is called π, which is a special number that is approximately equal to 3.14, or $\frac{22}{7}$.

Ask the students to tell you in their own words how the diameter relates to the radius of a circle.

Error Analysis

In Exercises 2 and 3, students may make errors in multiplying fractions. Remind them that, to multiply a whole number by fraction, they can write the whole number as a fraction with a denominator of 1. In Exercise 3, students may forget the 2 in the formula. Their work should look like this:

$$C = \frac{2}{1} \times \frac{22}{7} \times \frac{7}{1} = \frac{2}{1} \times \frac{22}{1} \times \frac{1}{1} = 44$$

Emphasize that 3.14 and $\frac{22}{7}$ are approximations of π and that all answers obtained when pi is expressed in either form are estimates.

Follow-Up

CROSS-CURRICULAR CONNECTION: HISTORY Ask students to find and share some information on the history of π. They will find that many ancient civilizations knew about the ratio of a circumference to the diameter of a circle.

LESSON 13

Introducing the Lesson

Draw a rectangle on the board. Label it 3 ft by 2 ft. Tell the students that this is the plan of a floor space that will be covered with one-foot square tiles. Draw grid lines to show the area divided into 3 foot parts one way and 2 foot parts the other way so that all 6 square feet are outlined.

Teaching the Lesson

Draw a rectangle on the chalkboard and shade the area inside it. Tell students that the area of the rectangle is this shaded region. Point out that the region can be measured in square units, such as square inches, square feet, and so on.

Ask students to describe real-life situations in which they would need to find the area of a rectangular region. They may mention carpeting, fertilizer or grass seed, or wall paint. Encourage them to name as many situations as they can.

Before they do the Application, it may help students to see a three-dimensional model. Show

LESSON PLANS

them any size box that has all six faces. Discuss the area of each face and how to find the total square inches of wrapping paper needed to cover the box.

Error Analysis

Some students may confuse the formulas for area and perimeter of a square. For Exercise 5, they may write 32 square centimeters. Have them draw the square on grid paper. Then have them count the square grid units inside the square, pointing out the 8 rows of 8 small squares.

Follow-Up

CONSUMER CONNECTION Have students find the number of square feet of wall space that must be painted if every wall of the classroom gets paint. Tell them to use round measurements to the nearest foot. Remind them that the windows don't get painted.

LESSON 14

Introducing the Lesson

On the chalkboard, draw a rectangle, square, parallelogram, and trapezoid. Ask students to tell you how the shapes are alike (four sides) and how they are different. Introduce the words *parallelogram* and *trapezoid,* if necessary. Relate the word *parallelogram* to the idea of parallel sides.

Teaching the Lesson

Have students use grid paper to draw some different parallelograms. On the chalkboard, show them how to remove a triangle from one side and place it on the other to form a rectangle with the same area. Have them determine the base of the parallelogram and the height of the parallelogram and count the square units to find the area. Then, read the area formula given in the lesson.

Ask students if the formula for the area of a parallelogram can be used for the area of a rectangle. If so, what would b and h be for the rectangle? (yes; b and h would be the length and width)

It may help students to set aside pages in their notebooks to summarize all of the geometry formulas they have used in the book so far. Suggest they make a diagram for each formula: $p = 2l + 2w$, $p = 4s$, $c = \pi d$ or $c = 2\pi r$, $A = lw,$ and $A = bh.$

Error Analysis

In Exercise 3, some students may get the result 306.0. This means that they have forgotten how to multiply decimals. Remind them that the number of places in the product is the sum of the number of places in each factor. Guide students to estimate their answers first as a way of checking the reasonableness of the answer they compute. Remind students that area is expressed in square units.

Follow-Up

FAMILY MATH Encourage students who sometimes care for younger children to paste some grid paper on cardboard, draw some parallelograms of different sizes and cut off a right triangle from each, as shown in the lesson. They can mix up all the pieces and take turns with the child finding matching pieces. Have the child make a rectangle and a parallelogram with the matching pieces.

LESSON 15

Introducing the Lesson

Ask students to name real objects that have the shape of a triangle. Some examples might be bridge trestles, tent openings, or pieces of pie. Encourage students to name as many different examples as they can.

Teaching the Lesson

Have students draw several different types of triangles on grid paper, leaving lots of room around each one. Show them how to double the triangle so that a parallelogram is formed. Point out that the area of the triangle must be half of the area of the parallelogram they drew. Ask them if the base and height of the triangle are the same as the base and height of the parallelogram (yes). Ask: What can you say about the size relationship between a trangle and a parallelogram that have the same base and height? Then have them read the example in the lesson.

Point out that any side of the triangle can be a base of the triangle. The height will always be perpendicular to the base.

When the area is not a whole number, it can be written using fractions or decimals, whichever the student prefers. For example, for Exercise 3, the result can be written 17.5 cm^2 or $17\frac{1}{2}$ cm^2. Remind students to include the correct label on their answers and that answers should be in square units.

Error Analysis

When three measures of a triangle are given, some students may not use the correct measures in the formula. Have students who make this error draw the triangle on grid paper with a horizontal base. This activity will help them recognize the correct base and height for a triangle.

Follow-Up

REAL-LIFE CONNECTION Have students do this experiment with straws: Tape together three lengths of a straw to form a triangle. Then tape together four lengths of a straw to form any four-sided shape. Try to change the angles in the triangle, then the quadrilateral. Students should notice that the triangle remains the same shape, but the quadrilateral can change shape; its angles can take different measures. Point out that this quality is what makes triangles useful in building design.

LESSON 16

Introducing the Lesson

Ask students about buying pizza locally, using a question like this: How large is a large pizza? What is the difference in price between large, medium, and small? Which is the best buy? How do you know?

Teaching the Lesson

Have students draw a circle on grid paper. Have them work in pairs to find the area of the circle by counting whole and partial square units. Have them also determine the radius in unit lengths.

Write on the chalkboard the area formula for circles, $A = \pi r^2$. Have the students compute the area of the circles they drew given the radius in grid lengths. Then have them compare their original answers with the result they get using the formula.

Students may need practice finding the square of a number. In Exercise 6, the radius is a mixed number or decimal, $1\frac{1}{2}$ or 1.5. Remind students how to multiply mixed numbers and how to multiply decimals.

Error Analysis

Some students may multiply the radius by 3.14 and then square the result. These students are ignoring the rules for order of operations. Review order of operations by discussing examples like $3(5)^2$, $(3 \times 5)^2$, $1 + 2^2$ and $(1 + 2)^2$. Guide students to recognize the difference between πr^2 and $(\pi r)^2$.

Follow-Up

REAL-LIFE CONNECTION A public park may have a fountain, statue, or flower garden in a circular area. Often, this circular area is surrounded by a walkway. If there is such an arrangement in your locale, mention it or ask students to describe any such arrangement they have seen. Draw two concentric circles on the chalkboard. Shade the area that would be the circular walkway. Ask students to write a formula for the area of the walkway only, using R for the radius of the larger circle and r for the radius of the smaller circle. Alternately, you could give measures for each radius and have the students compute the area.

LESSON PLANS

LESSON 17

Introducing the Lesson

Bring in two different sizes of cereal boxes. Tell the students you want to know which box holds more. Ask them if comparing any one dimension—length, width, or height—will tell them which holds more. Ask them to explain their answers.

Teaching the Lesson

If possible, show the students 3-dimensional models of cubic inches or cubic centimeters. Draw a diagram on the chalkboard of a box that measures 3 units by 2 units by 4 units. With the cubic unit models, make one layer of units, 3 units by 2 units, to represent what will fit in the bottom of the box. Have students verify that 6 cubic units fit along the bottom of the box. Then, add 3 additional layers and have students verify that 6×4 or 24 cubic units fit in the box. Write the formula $V = lwh$ on the chalkboard and relate it to what you did with the cubic units.

Ask students to explain why comparing volumes will tell them which of the two boxes contains more cereal.

When discussing the volume of a cube, point out that the volume is still the product of all three dimensions, but that a different symbolism can be used since all three dimensions have the same measure.

Another formula that works for the volume of both of these shapes is $V = Bh$ where B is the area of the base and h is the height. You may wish to introduce this formula to students.

Error Analysis

In the Application, students may try to guess which box holds more, based on appearance. Even though the Rice Pilaf box is taller and broader, the Basmati Rice box holds more. Point out that looks can be deceiving and that, to compare volumes, they really must compute the volume for each box.

Follow-Up

REAL-LIFE CONNECTION Have students compute the volume of the room they are in. Ask: How many cubic inches of air fill the room? How many cubic inches of air are there for each person in the room?

LESSON 18

Introducing the Lesson

Ask students to describe the shape of a can as if to someone who has never seen one. After they have identified that a circle relates to the shape of a can, ask them to recall the formula for computing the area of a circle.

Teaching the Lesson

Before computing the volumes of the two cans, ask students to identify and compare the dimensions of the cans. The chicken soup can is taller and the vegetable soup can is wider. When the volumes have been computed, ask students if they think the taller can will always have the greater volume. Have them explain their reasoning.

Remind students to be careful when evaluating the formula. They must use the rules for order of operations and find the square of the radius before multiplying by π and the height.

Notice that, as in the previous lesson, the formula $V = Bh$ works, where B is the area of the base and h is the height. This general formula may be easier for students to remember. Challenge students to explain how they would determine which of two different-size cans of pet food is the better buy.

Error Analysis

For students who forget to use the radius when the diameter of the can is given, have them develop the habit of underlining the r in the formula. You may also want them to draw a radius on whatever figure is given and label it.

12 FORMULAS

LESSON PLANS

Follow-Up

CAREER CONNECTION Explain that package designers are employed by many manufacturing companies. Ask students to list the things a package designer would have to consider when deciding on the best size and shape of a can to use for a particular product.

LESSON 19

Introducing the Lesson

Ask students to describe the shape of a pyramid. Have them tell what they know of the pyramids of Egypt, Mexico, and Central America. Ask them if they know of other real-life examples of pyramids.

Teaching the Lesson

Some students may be surprised to know that pyramids can have different shapes for their bases. Draw attention to the definition of a pyramid provided in the Vocabulary section of the lesson.

Instead of presenting a different formula for pyramids with different bases (the formula presented), point out that $V = \left(\frac{1}{3}\right)Bh$ applies to all pyramids. To use the formula in a particular case, students must replace B with the formula for the area of that polygon.

Be sure students realize that the height of the pyramid is measured by a perpendicular line segment connecting the vertex to the base of the pyramid. Emphasize that it is NOT the height of one of the triangular faces of the pyramid.

Error Analysis

For Exercises 2–5, students must recognize the shape of the base of the pyramid, and use the correct dimensions and the correct area formula in the formula for the volume. Exercises 3 and 4 have triangular bases. Students may not remember that they need to know the base and height, which are perpendicular, to find the area of the triangle. Some students may forget that the area of a triangle is one-half the base times the height. For these students, review the area formulas.

Follow-Up

CROSS-CURRICULAR CONNECTION: HISTORY Students may be interested to know that the Rhind papyrus, copied by the scribe Ahmes in about the year 1650 B.C., showed that the Egyptians knew that the volume of a square pyramid was $\frac{1}{3}$ the volume of a rectangular prism having the same base and height. The Rhind papyrus is a collection of mathematical examples from ancient Egypt. It was bought in a shop in Luxor by a Scottish scholar, A. Henry Rhind, in 1858 A.D., and donated to the British Museum after his death.

Challenge students to design a way to prove that when the area of the bases are the same and when the heights are the same, a prism will have 3 times the volume of a pyramid.

LESSON 20

Introducing the Lesson

Ask students to describe the shape of a cone and tell you what objects they have seen that have this shape. They may mention ice cream cones, drinking cups, or megaphones. Ask students to predict the relationship between the volumes of a cylinder and a cone when both have the same height and bases with equal areas.

Teaching the Lesson

Ask students to tell you how a cone is like a cylinder and how it is different. The relationship between the volume of a cone and the volume of a cylinder with the same height and base is the same as the relationship between the volume of a pyramid and the volume of a prism with the same height and base area. The relationship is that one volume is one-third of the other.

LESSON PLANS

As with the formula for the volume of a cylinder, students must be careful to use the radius and not the diameter in this formula.

Ask students if multiplying by $\frac{1}{3}$ is the same as dividing by 3. Some students may prefer to use the formula $V = \frac{\pi r^2 h}{3}$ for finding the volume of a cone.

Error Analysis
In Exercise 4, if students get the result 1365 cm³, they used the diameter instead of the radius. Have them draw a new diagram on which they have the height labeled 16 cm and the radius labeled 8 cm.

Follow-Up
FAMILY MATH Encourage students who sometimes care for younger children to play with sand or salt and compare volumes of small cans, cones, and different sizes of boxes. Also, it's interesting to have family members speculate on whether a short, wide mug or glass holds more than a tall, thin one. The results are often surprising. You may wish to point out that many young children are not yet able to observe volume, and will invariably choose the taller container. Warn students that they may not be able to convince the younger child that he or she is incorrect.

LESSON 21

Introducing the Lesson
Ask students if they can recall a time when they were someplace that the temperature was below zero. Ask them to describe how a temperature below zero is written.

Teaching the Lesson
Draw a vertical number line on the chalkboard. Mark 55° and −8°. Help students to see that the difference between these two points is 55 + 8 or 63.

Then draw a horizontal number line. Point out that positive numbers always appear to the right of 0 on the number line and negative numbers always appear to the left. The numbers increase in value from left to right. Ask volunteers to take turns finding pairs of numbers that are 12 units apart on the number line. Encourage students to include some pairs that have opposite signs.

Write $|4|$ and $|-3|$ on the chalkboard. Point out that the symbol means absolute value. The absolute value of a number is its distance from zero on the number line. Since absolute value is a number of units, it is always a positive number.

$$|4| = 4 \qquad |-3| = 3$$

Ask students to name two points that have an absolute value of 5. Ask them to tell which number has a greater absolute value, 12 or −12.

The distance between two points is defined as the absolute value of the difference between them. Therefore, for 4 and 7, the distance between them is $|4 - 7| = |-3| = 3$ or $|7 - 4| = |3| = 3$.

Error Analysis
When using absolute value, some students will recognize that $|-2| = 2$ but will want to say that $|2| = -2$. In other words, they will interpret absolute value as the opposite of the number all of the time since it does equal the opposite of the number some of the time. Emphasize that the absolute value of a number is the number of units it is from 0. Show that −8 has a greater absolute value than +3.

Follow-Up
REAL-LIFE CONNECTION Another situation that can be modeled using a number line is distance above and below sea level. Ask students to use an almanac or other source to find the difference in elevation between the highest and lowest points in the United States. [The elevation at Mt. McKinley, Alaska, is 6,198 meters above sea level and at Death Valley, California is 86 meters below sea level. (6284 m)]

14 FORMULAS

LESSON PLANS

LESSON 22

Introducing the Lesson

Ask students to recall road or street maps they have used. Ask: How did the map help you locate a particular place? Point out that a street map is a coordinate system. Encourage discussion of different types of coordinate systems.

Teaching the Lesson

Draw a horizontal number line. Mark coordinates to the left and right of 0 and label them. Remind students that positive numbers are to the right, negative numbers to the left. Then, draw a vertical number line perpendicular to the horizontal one and through the 0 point. Mark coordinates up and down from 0 and label them. Point out that positive numbers are above the horizontal axis and negative numbers are below the horizontal axis. As students read the lesson, label the horizontal number line x and the vertical number line y.

On the coordinate plane you drew on the board, ask volunteers to come and locate points $(-1, 4)$, $(2, -4)$, $(-3, -2)$ and $(2, 1)$.

It is important to show that $(-2, 1)$ is different from $(1, -2)$. This is why the pairs are called ordered pairs.

Error Analysis

Some students may confuse the x- and y-coordinates when graphing or reading ordered pairs. They should always label the axes when they draw them. Then, tell them that, since x comes before y in the alphabet, the first coordinate is always the x-coordinate. Emphasize that the x-coordinate always is a measure of the distance from the y-axis and that the y-coordinate always gives the distance from the x-axis.

Follow-Up

CAREER CONNECTION Have students discuss the career of marketing. Businesses want to make decisions based on facts and what they can reasonably infer from facts. One useful tool is graphing. Real data can be expressed as ordered pairs and then graphed. The graphed points can be examined for patterns. For example, to understand the relationship between the price of an item and the number sold, a company analyst might gather information in the form (price, number sold) and graph it. The graph may indicate the best price for an item.

LESSON 23

Introducing the Lesson

Display two boards or sheets of cardboard. Have them both meet the top of the desk at an angle so that they each form a kind of ramp, but make one angle greater than the other. Ask students which ramp they think a ball will roll down faster. Have them explain their choice and describe how the faster ramp is different from the slower ramp.

Teaching the Lesson

Explain that slope is the steepness of a curve. Since the steepness is the same everywhere, it can be measured between any two points. Draw a pair of coordinate axes on the chalkboard. Label one point (x_1, y_1). Point out that you are using subscripts to show that this is a particular point. Then, label another point on the graph (x_2, y_2). Point out that the subscripts just mean that this a different particular point on the line.

Draw a dashed vertical line from (x_2, y_2) to the position that corresponds to y_1. Then, from that position, draw a dashed horizontal line to the point (x_1, y_1). Ask students to give an expression for the vertical distance $(y_2 - y_1)$. Then, ask them to give you an expression for the horizontal distance $(x_2 - x_1)$. Write the formula $slope = \frac{(y_2 - y_1)}{(x_2 - x_1)}$. Guide students to see that slope is the change in y-coordinates divided by the change in x-coordinates.

LESSON PLANS

To develop the slope of horizontal and vertical lines, give specific examples and use the slope formula. For a horizontal line, you can use (−3, 2) and (4, 2). For a vertical line, you can use (2, 5) and (2, −1).

Some students may want to know why division by zero is undefined. Explain using related sentences. For example, $14 \div 7 = 2$ since $7 \times 2 = 14$ and $0 \div 3 = 0$ since $3 \times 0 = 0$. However, $2 \div 0$ means the same as $0x? = 2$. But any number multiplied by 0 gives 0, not 2. Therefore, division by 0 is not defined.

Error Analysis
Some students will mix up x- and y- coordinates when using the slope formula. For example, for Exercise 2 they will get $\frac{2}{3}$ instead of $\frac{-2}{3}$ because they will compute $\frac{(-1-1)}{(-2-1)}$. Have students who make this error circle the ordered pair of one point and always use that point's coordinate first in the formula.

Students who get $\frac{3}{2}$ for Exercise 2 are placing the difference of the x-coordinates in the numerator and the difference of the y-coordinates in the denominator. Have them write the formula on their papers while they are doing the exercises.

Follow-Up
REAL-LIFE CONNECTION Draw three graphs on the chalkboard, the first with a positive slope, the second with a negative slope, and the third a horizontal line. Tell students that these three graphs show the relationship between temperature and time in three different cities. Ask which graph shows a decrease in temperature, which shows no change in temperature, and which shows an increase in temperature. (negative, horizontal, positive)

LESSON 24

Introducing the Lesson
Ask students to describe a right triangle. Have volunteers draw several different right triangles on the chalkboard.

Teaching the Lesson
For the right triangles drawn on the board, label each leg and hypotenuse. Point out that the hypotenuse is always the side across from the 90° angle. Ask: How does the length of a hypotenuse compare in size with the length of the other sides of a right triangle? (It is always the longest side.)

Remind students that a^2 means $a \times a$. If many of the students do not have calculators, you may want to display a table of squares for the numbers 1 through 26.

When students have completed the Guided Practice, have them verify that $6^2 + 8^2 = 10^2$.

Error Analysis
Some students may always add the squares of the two given measures, regardless of which measures have been given. Remind students of the theorem. Tell them to write the theorem, treat it as a formula, substitute the correct data, and solve the equation for the unknown.

Follow-Up
CROSS CURRICULAR CONNECTION: HISTORY Tell students that the ancient Egyptians used a circle of rope divided into twelve equal parts to mark off a square corner for a building. They placed a stake in the ground, marked off three units and placed another stake, then marked off four units and placed another stake so that the rope was taut and formed a triangle. Ask students to explain why this system gave a square corner. (They made a 3-4-5 triangle.)

LESSON 25

Introducing the Lesson
Draw a right triangle on the chalkboard. Label one leg 9 in. and the other leg 12 in. Ask students to tell you how to find the length of the hypotenuse.

Teaching the Lesson

Draw a set of coordinate axes on the chalkboard. Label two points diagonally distant from one another (x_1, y_1) and (x_2, y_2) and draw a line segment between them. Then, show the line segment as the hypotenuse of a right triangle by drawing dashed legs dashed vertical and horizontal lines, as you did in the slope lesson. Ask students what figure is formed by the three line segments. Ask them how they can use what they learned about the Pythagorean theorem to help them find the length of the line segment that is the distance between the two points.

While the formula looks very complicated, it is a direct application of the Pythagorean theorem and so can always be thought through rather than memorized. Point out that each diagonal segment can be the hypotenuse for two right triangles. Students can see that when the horizontal and vertical sides are drawn, the hypotenuse is the diagonal of a rectangle.

Point out that it does not matter which point is called (x_1, y_1) and which is called (x_2, y_2) as long as each x-coordinate is matched with the corresponding y-coordinate.

Error Analysis

Students may make errors in subtracting signed numbers. Suggest that these students draw number lines to find the differences.

Follow-Up

FAMILY MATH Encourage students who sometimes care for younger children to play a game with a coordinate grid, drawn on grid paper or construction paper. (For younger children, the grid can only have positive numbers on each axis.) Make these suggestions: Take turns following directions like: Go up 3 and over 2 to the right. Then, mark two points on the grid and ask the child to find the shortest route between the two points. Let the child choose two points and have you find the shortest route between them.

ANSWER KEY

LESSON 1 (pp. 2–3)

1. **a.** the number of quarters in Tyrone's pocket
   b. q (letters may vary); the number of quarters in Tyrone's pocket
   c. $.25, or 25¢
   d. multiply
   e. $.25q$, or $25q$

   NOTE: Any letter may be used for a variable for Exercises 2–7.

2. $5.50h$; h represents the number of hours worked

3. $t - $.75$; t represents the original price of the toothpaste

4. $.89b$; b represents the number of pounds of bananas bought

5. Sample: The amount of money earned, if you earn $10 per hour, and h represents the number of hours worked.
   Sample: The number of people at a party, if 6 people leave, and f represents the number of people at the party before the 6 left.

6. $3p$; p represents the price of one CD

7. Sample: The total number of people in line at a cash machine, if m represents the number of males in line, and f represents the number of females in line.

8. Sample: The amount of money earned, if you earn $10 per hour, and h represents the number of hours worked.

LESSON 2 (pp. 4–5)

1. Add inside parentheses.
   a. $6(5 + 3) - 3 \div 1$
   Multiply. **b.** $6(8) - 3 \div 1$
   Divide. **c.** $48 - 3 \div 1$
   Subtract. **d.** $48 - 3$
   e. 45

2. 6
3. 483
4. 26
5. 120
6. 18
7. 6
8. Answers may vary.
9. 4.5
10. Answers may vary. Sample: If we all use different rules, then we will all get different answers.

LESSON 3 (pp. 6–7)

1. **a.** $6(2 + 10) - 3$ **b.** 12 **c.** 72 **d.** 69
2. **a.** the number of quarters; the number of dimes
   b. $.25 \times 5 + $.10 \times 4$ **c.** $1.25, $.40
   d. $1.65
3. $298
4. 43
5. $2\frac{1}{8}$
6. $33
7. $61.50
8. $19.25
9. 29
10. $31
11. Sample: Carl bought three items priced at $4.99 and two items priced at $2.29. The total cost was $19.55, which is the value of the expression.

LESSON 4 (pp. 8–9)

1. **a.** subtraction
   b. $n = 14$
2. **a.** addition
   b. $n = 25$
3. **a.** division
   b. $n = 8$
4. **a.** division; multiplication
   b. $n = 45$
5. $a = 33$
6. $y = 5
7. $n = 4$
8. $h = 14$
9. $y = 2.00
10. $a = 16

ANSWER KEY

11. a. $x + 4 = 12$
b. 8

LESSON 5 (pp. 10–11)
1. a. centimeters
b. multiplication; division
c. 100
2. $t = d \div r$
3. $m = F \div a$
4. $p = i \div (r \times t)$
5. $t = p - g$
6. $d = g - p$
7. $h = \frac{2a}{b}$
8. 9 inches
9. $b = \frac{20}{h}$
10. Sample: In Exercise 8, all variables are given. In Exercise 9, the height is not given so you solve for the base in terms of the height.

LESSON 6 (pp. 12–13)
1. a. distance
b. $r = 65, t = 3$
c. $d = 195$; 195 miles
2. a. time
b. $d = 300; r = 60$
c. $t = d \div r$
d. $t = 5$; 5 hours
3. 130 miles
4. about 373 mi/hr
5. about 9 hours
6. a. $4\frac{2}{3}$ hours
b. by car
7. $165.3 \div r$

LESSON 7 (pp. 14–15)
1. a. 10
b. 18
c. 50°
2. a. 50°
b. $\frac{5}{9}$
c. 10°
3. 100°C

4. 98.6°F
5. 20°
6. a. 5°C
b. 22°C
c. 35°C
7. a. 38°F [exact value 37.4°F]
b. 62°F [exact value 59°F]
c. 132°F [exact value 122°F]
8. Fahrenheit to Celsius: The strategy works because 30 is about 32, and dividing it by 2 is about the same as multiplying by $\frac{5}{9}$. Celsius to Fahrenheit: The strategy works because $\frac{9}{5}$ is about 2.

LESSON 8 (pp. 16–17)
1. a. 90, 500, 3
b. $\frac{I}{pt}$
c. 90, 500, 3
d. 90; 1,500; 0.06
e. 6%
2. $400
3. $50
4. $1,125
5. 7%
6. $1,045
7. about 5.6%
8. 20 years old

LESSON 9 (pp. 18–19)
1. a. Sales tax = rate × price
b. 0.06
c. $2.10
d. Total cost = price + sales tax
e. $37.10
2. $21.20
3. $29.40
4. $3.07
5. $20.87
6. $15.44
7. $25.41
8. $12.78
9. $6.33
10. 7%

ANSWER KEY

11. Enrique adds 1 to the sales tax rate and multiplies the price by 1.05, instead of multiplying to find the sales tax and then adding that to the price to get the total cost.

LESSON 10 (pp. 20–21)

1. **a.** discount = rate × price
   b. 0.15
   c. $7.20
   d. total cost = price − discount
   e. $40.80
2. **a.** $29.98 × $\frac{1}{2}$ = $14.99
   b. $14.99
3. $42
4. $90
5. $296.10
6. $748
7. $126.75
8. $15.50
9. $8
10. $11
11. Casual Dress; discounted price is $33 at Buy Jeans, $34 at Blue Denim, $29.33 at Casual Dress, $29.99 at Jeans, Etc.

LESSON 11 (pp. 22–23)

1. **a.** 4
   b. 12 units, 5 units, 12 units, 5 units
   c. 34 units
2. **a.** 3
   b. 15 cm, 15 cm, 15 cm
   c. 45 cm
3. 20 yd
4. 20 in.
5. 30 cm
6. 32 m
7. 36 units
8. 24 m
9. 58 units
10. 100 cm or 1 m
11. $P = s + s + s + s$ or $P = 4s$
12. $P = l + w + l + w$ or $P = 2l + 2w$
13. $P = a + b + c$
14. 55 ft, 4 in.

LESSON 12 (pp. 24–25)

1. **a.** diameter
   b. $C = \pi d$
   c. 3.14 or $\frac{22}{7}$
   d. about 16 cm
2. 22 in.
3. 44 ft
4. 31.4 yd
5. 25.12 m
6. about 12 ft
7. 3.2 cm
8. 28 m
9. 35 cm
10. Answers may vary. Sample: The diameter of a circle is twice as long as its radius ($D = 2r$). So, $\pi D (2r) = 2\pi r$.
11. Check students' work.
12. The diameter of the second circle will be twice the length of the diameter of the first circle.

LESSON 13 (pp. 26–27)

1. **a.** 5 × 5 square **b.** 25 square units
   c. Area = side × side or $A = s^2$
   d. 25 units; yes
2. 24 sq units
3. 108 m^2
4. 4,700 ft^2
5. 2,808 sq ft
6. 225 in.^{2}
7. 6,400 yd^2
8. **a.** 13.5 cm^2
   b. $49\frac{1}{4}$ in.^{2}

LESSON 14 (pp. 28–29)

1. **a.** 8 units
   b. 3 units
   c. $A = bh$
   d. 24 square units
2. 40 in.^{2}
3. 30.6 m^2
4. 108 cm^2

ANSWER KEY

5. $44\frac{1}{3}$ yd^2
6. 16 sq units
7. 6 sq units
8. Because the width is perpendicular to the length, the width of a rectangle can be called the height. So the formula Area = length × width can be written as Area = base × height, which is the area of a parallelogram. So, the formulas are the same. A parallelogram and a rectangle with the same base and height cover the same amount of surface.

LESSON 15 (pp. 30–31)

1. a. 10 cm
   b. 20 cm
   c. $A = \frac{1}{2}bh$
   d. 100 cm^2
2. 24 in.^{2}
3. 17.5 cm^2
4. 30 in.^{2}
5. 22 m^2
6. 96 in.^{2}
7. 12.5 ft^2
8. 10 square units
9. 6 square units
10. Yes, because the dark and the light areas cover the whole block, 16 sq units.

LESSON 16 (pp. 32–33)

1. a. $15 \div 2 = 7\frac{1}{2}$ in.
   b. $A = \pi r^2$
   c. 176.625 in.^{2}
2. 28.26 m^2
3. 314 cm^2
4. 12.56 ft^2
5. 35,949.86 yd^2
6. 7.065 sq in.
7. 254.34 ft^2
8. jumbo
9. The area of a circle (pizza) is the amount of surface. The price per square unit of surface lets you compare the cost of the pizzas.

LESSON 17 (pp. 34–35)

1. a. 8 cm
   b. width = 8 cm, height = 8 cm
   c. $V = lwh$
   d. 512 cu cm
2. 1,774.5 cu cm
3. 1,512.5 cu cm
4. 125 cu cm
5. $225\frac{5}{8}$ cu in.
6. $42\frac{7}{8}$
7. Basmati Rice box
8. Samples: the amount of product the box can hold (volume), the amount of cardboard needed to make the box (surface area), the amount of shelf space required, how the box looks, room available for advertising.

LESSON 18 (pp. 36–37)

1. a. 4 cm
   b. $\pi = 3.14, r = 4$ cm, $h = 8.5$ cm
   c. $V = \pi r^2 h$
   d. 427.04 cu cm
   e. 427 cu cm
2. 176 cu cm
3. 39 cu in.
4. 305 cu cm
5. 145 cu cm
6. the taller canister
7. No, because the cookies might fit better into one canister than the other, leaving less space.

LESSON 19 (pp. 38–39)

1. a. $w = 5$ ft, $h = 8$ ft
   b. $V = \frac{1}{3}Bh$
   c. $6 \times 5 = 30$ sq ft
   d. 80 cu ft
2. 8 cu in.
3. 72 cu cm
4. 453 cu ft
5. 820 cu cm
6. 267 cu m
7. 3,008 cu m
8. 195 cu in.

ANSWER KEY

LESSON 20 (pp. 40–41)
1. a. $r = 4$ m, $h = 6$ m
   b. $V = \frac{1}{3}\pi r^2 h$
   c. 100.48 cu m
2. 1,780 cu m
3. 1,047 cu ft
4. 1,072 cu cm
5. 64 cu in.
6. 804 cu cm
7. 196 cu cm
8. 2,093 cu ft
9. 144 cu m
10. 16 cu m
11. The volume of a cone is one-third the volume of a cylinder with the same radius and height.

LESSON 21 (pp. 42–43)
1. a. -3
   b. 2
   c. $|-3 - 2| = |-5|$
   d. 5
2. 3
3. 2
4. 10
5. 6
6. 8
7. 8
8. 8
9. 19
10. Distance between A and B is $|a - b|$. If the coordinate of B is zero, then $|a - b| = |a - 0| = |a|$.

LESSON 22 (pp. 44–45)
1. a. (4, 3)
   b. (2, −3)
   c. (−3, −2)
   d. (−4, 1)
2. H
3. K
4. M
5. N
6. G
7. F
8. (2, 4)
9. (0, 0)
10. (−2, 2)
11. (4, −2)
12. (−4, −3)
13. $(2\frac{1}{2}, 0)$
14. Drawings will vary. Check that points are labeled with the correct coordinates.

LESSON 23 (pp. 46–47)
1. a. fall
   b. 5 units
   c. −2 units
   d. $-\frac{2}{5}$
2. a. 5
   b. −2
   c. $-\frac{2}{5}$
3. $P(-2, 1)$, $Q(1, -1)$, slope $= -\frac{2}{3}$
4. $P(2, 2)$, $Q(1, -2)$, slope $= 4$
5. D
6. C
7. A
8. B

LESSON 24 (pp. 48–49)
1. a. hypotenuse
   b. $b = 8$
   c. 36, 64
   d. 36 + 64 or 100
   e. 10
   f. 10 city blocks
2. 5 ft
3. 13 in.
4. 9 cm
5. 17 m
6. 8 in.
7. 26 mm
8. 50 cm
9. 7 ft
10. no, because $3^2 + 5^2 = 9 + 25 = 34$ and $7^2 = 49$

23

ANSWER KEY

11. no, because $1^2 + 13^2 = 1 + 169 = 170$ and $14^2 = 196$

LESSON 25 (pp. 50–51)

1. **a.** Choices will vary.
   b. 4, 3
   c. 16, 9
   d. 25
   e. 5
2. 5
3. 10
4. In $\sqrt{(a^2 + b^2)}$, find the squares first, then add, then find the square root.

CUMULATIVE REVIEW (L1–L4) (p. 52)

NOTE: In exercises 1–4, any letter may be used for a variable.

1. $4a$; a represents the price of each adult ticket.
2. $.69p$; p represents the number of pounds of apples.
3. $6 + 1 + p$ (or $7 + p$); p represents the number of people in line behind Bill.
4. $.25q + $.10d$; q represents the number of quarters and d represents the number of dimes.
5. 8
6. 51
7. 44
8. 5
9. 45
10. $3.31
11. $250
12. $20
13. $p = 8$
14. $y = 27$
15. $n = 7$
16. $m = 214$

CUMULATIVE REVIEW (L5–L7) (p. 53)

1. $a = F \div m$
2. $r = I \div p \div t$, or $r = \frac{I}{pt}$
3. $t = p - g$
4. $s = P \div 4$
5. 240 miles
6. 332.5 mph
7. $14\frac{1}{2}$ hours
8. $146\frac{1}{4}$ miles
9. 45°C
10. 60.8°F
11. 5°C
12. 212°F

CUMULATIVE REVIEW (L8–L10) (p. 54)

1. $75
2. 6%
3. $313.50
4. $44.52
5. $31.48
6. $24.31
7. $40
8. $18.70
9. 25%
10. $509.85

CUMULATIVE REVIEW (L11–L16) (p. 55)

1. perimeter = 22 in.; area = 28 sq in.
2. perimeter = 20 ft; area = 25 sq ft
3. perimeter = 40 cm; area = 60 sq cm
4. perimeter = 40 m; area = 40 sq m
5. circumference = 6.28 ft; area = 3.14 sq ft
6. circumference = 37.68 cm; area = 113.04 sq cm
7. perimeter = 44 ft; area = 121 sq ft
8. perimeter = 12 m; area = 6 sq m
9. perimeter = 30 in.; area = 54 sq in.
10. circumference = 25.12 cm; area = 50.24 sq cm

ANSWER KEY

CUMULATIVE REVIEW (L17–L20) (p. 56)

1. 180 cu in.
2. 343 cu cm
3. 1,060 cu cm
4. 1,005 cu in.
5. 72 cu ft
6. 448 cu in.
7. 324 cu in.
8. 451 cu cm
9. 135 cu m
10. 90 cu cm

CUMULATIVE REVIEW (L21–L25) (p. 57)

1. 6
2. 4
3. 10
4. 6
5. 13 ft
6. 8 in.
7. **a.** $P(-4,4)$, $Q(4,-2)$
   b. $-\frac{3}{4}$ **c.** 10
8. **a.** $P(-3, 1)$, $Q(3, 1)$
   b. 0 **c.** 6
9. **a.** $P(-2,2)$, $Q(1,2)$
   b. 0 **c.** 5
10. **a.** $P(-1, -2)$, $Q(2, 2)$
    b. $\frac{4}{3}$ **c.** 5

ACCESS TO MATH
can help you work on a wide range
of math skills.

Look for these titles

- WHOLE NUMBERS AND INTEGERS
- FRACTIONS
- RATIOS AND PROPORTIONS
- DECIMALS
- PERCENTS
- MEASUREMENT AND GEOMETRY
- ESTIMATION
- GRAPHING AND INTERPRETING DATA
- BASIC STATISTICS
- PROBABILITY
- FORMULAS
- EXPONENTS AND SCIENTIFIC NOTATION
- PROBLEM SOLVING STRATEGIES
- ONE- AND TWO-STEP PROBLEMS
- PATTERNS AND FUNCTIONS

GLOBE FEARON EDUCATIONAL PUBLISHER
A Division of Simon & Schuster
Upper Saddle River, New Jersey

ISBN 0-8359-1588-3

90000

9 780835 915885

Y0-AWK-160